The Art of Being Still: A Guide to the Subtle Mastery of Doing Nothing

Scriptum Books

Table Of Contents

The Myth of Busyness: Unraveling the Cult of Productivity 9
- **The Origins of Busyness** 9
- **The Illusion of Productivity** 10
- **Breaking Free from the Cult of Productivity** 10
- **Cultivating a Culture of Rest** 10
- **Embracing the Slow Movement** 11

Embracing Stillness: A Journey into the Heart of Silence 12
- **The Essence of Stillness** 12
- **The Path to Silence** 13
- **The Dance of Silence and Sound** 13
- **Finding Refuge in Silence** 13
- **The Wisdom of Silence** 14
- **Cultivating Stillness in Daily Life** 14

The Power of Presence: Cultivating Mindfulness in Everyday Life 16
- **Understanding Mindfulness** 16
- **The Benefits of Mindfulness** 17
- **Bringing Mindfulness into Daily Life** 17
- **Cultivating Presence in Relationships** 18
- **Overcoming Challenges** 18
- **Embracing Impermanence** 18
- **The Practice of Loving-Kindness** 19

Finding Peace in the Pause: Harnessing the Magic of Moments 21

- The Importance of Pausing 21
- The Practice of Mindful Pausing 22
- Embracing Stillness 22
- The Art of Doing Nothing 22
- Savoring the Present Moment 23
- Cultivating Presence in Relationships 23
- Overcoming Resistance 23
- Creating Rituals of Pause 24

Letting Go of Guilt: Liberating Yourself from the Need to Constantly Do 25
- The Guilt of Not Doing 26
- The Myth of Perpetual Productivity 26
- Embracing the Power of Rest 26
- Letting Go of Shoulds and Shouldn'ts 27
- Embracing Self-Compassion 27
- Rewriting the Narrative 28
- The Practice of Radical Rest 28
- Cultivating a Culture of Rest 28

The Joy of Solitude: Nurturing Your Relationship with Yourself 30
- The Beauty of Solitude 30
- Embracing Alone Time 31
- Cultivating Self-Companionship 31
- Finding Freedom in Solitude 32
- The Practice of Mindful Solitude 32
- Rediscovering Joy in Simple Pleasures 33

- Embracing the Wisdom of Silence 33

Rest as Resistance: Challenging the Tyranny of Constant Activity 35

- The Culture of Busyness 36
- The Myth of Perpetual Productivity 36
- The Power of Rest 36
- Restoring Balance 37
- The Practice of Radical Rest 37
- Cultivating a Culture of Rest 38

The Art of Doing Nothing: Discovering the Beauty in Non-Doing 39

- Redefining Productivity 40
- Embracing Stillness 40
- Surrendering to Non-Doing 40
- Cultivating Mindfulness 41
- Finding Joy in Simple Pleasures 41
- Letting Go of Guilt 42
- The Gift of Presence 42

Surrendering to Silence: Deepening Your Connection to Inner Wisdom 44

- The Wisdom of Silence 44
- Surrendering to Stillness 45
- Cultivating Inner Listening 45
- Embracing the Unknown 46
- Nurturing Self-Compassion 46
- Deepening Connection to Self and Others 46
- The Practice of Surrender 47

Nature's Serenade: Healing and Renewal in the Natural World 48
- **The Healing Power of Nature** 48
- **Reconnecting with Our Roots** 49
- **Awakening the Senses** 49
- **Finding Peace in the Wilderness** 50
- **Healing the Body, Mind, and Spirit** 50
- **Cultivating Gratitude and Reverence** 50
- **Embracing the Call of the Wild** 51

Mindful Movement: Exploring Stillness Through Yoga and Tai Chi ... 52
- **The Wisdom of Yoga and Tai Chi** 52
- **The Art of Yoga: Union of Body, Mind, and Spirit** 53
- **The Grace of Tai Chi: Cultivating Balance and Harmony** 53
- **Cultivating Mindfulness in Motion** 54
- **Embracing the Journey Within** 54
- **The Healing Power of Presence** 54
- **Integrating Mindful Movement into Daily Life** 55

The Beauty of Boredom: Embracing Creative Lulls and Restorative Rest 57
- **Rethinking Boredom** 57
- **Embracing Creative Lulls** 58
- **Cultivating Presence in the Moment** 58
- **Nurturing the Imagination** 59
- **Restorative Rest and Renewal** 59
- **Cultivating a Culture of Rest** 59

Sacred Spaces: Creating Sanctuaries for Stillness in Your Home and Mind 61

    The Importance of Sacred Spaces 61

    Creating Physical Sanctuaries in Your Home 62

    Designing Mental Sanctuaries in Your Mind 62

    Incorporating Ritual and Ceremony 63

    Cultivating Presence and Stillness 63

Letting Time Unfold: Trusting the Flow of Life's Rhythms 65

    Embracing the Present Moment 65

    Releasing the Grip of Control 66

    Trusting in Divine Timing 66

    Cultivating Patience and Resilience 67

    Finding Beauty in the Unfolding 67

    Living in Alignment with Life's Rhythms 68

The Slow Revolution: Rediscovering the Lost Art of Patience 69

The Wisdom of Non-Attachment: Releasing the Need for Constant Distraction 73

    Understanding Non-Attachment 74

    Letting Go of the Ego 74

    Embracing Impermanence 74

    Cultivating Contentment and Joy 75

    Letting Silence Speak 75

    Cultivating Non-Attachment in Daily Life 76

    The Liberation of Non-Attachment 76

The Practice of Deep Listening: Cultivating Presence in Relationships 78

    The Essence of Deep Listening 78

    Creating Space for Presence 79

    Cultivating Empathy and Understanding 79

    Honoring the Power of Silence 80

    Overcoming Barriers to Listening 80

    Deepening Connection and Trust 80

    Integrating Deep Listening into Daily Life 81

Restoring Balance: Nourishing Your Body, Mind, and Spirit 82

    The Importance of Balance 82

    Nourishing Your Body 83

    Cultivating Mental Well-being 83

    Connecting with Your Spirit 84

    Integrating Holistic Practices 84

    Embracing Self-Care 84

    Cultivating Gratitude and Joy 85

The Art of Savoring: Finding Delight in Life's Simple Pleasures 86

    The Essence of Savoring 87

    Engaging the Senses 87

    Cultivating Mindfulness 87

    Finding Joy in the Ordinary 88

    Creating Rituals of Savoring 88

    Embracing the Transience of Experience 88

    Sharing Savoring with Others 89

Embracing the Journey: Sustaining Your Practice of Being Still 91

The Continual Journey of Being Still..91

Cultivating Patience and Perseverance......................................92

Nurturing Self-Compassion and Kindness................................92

Embracing Resistance and Discomfort......................................93

Cultivating Presence in Everyday Life 93

Finding Support and Community 93

Celebrating Your Progress and Growth.....................................94

The best quotes about doing nothing 95

"So much leisure time, but what to do with it? Don't spend it staring into space—that's when uncomfortable thoughts surface and the void beckons."

— Maartje Willems

The Myth of Busyness: Unraveling the Cult of Productivity

In our modern society, busyness has become a badge of honor. We equate productivity with worth, and being constantly busy is often seen as a sign of success. But beneath the surface lies a pervasive myth—a belief that the more we do, the more valuable we are. In this chapter, we will explore the roots of this myth, its impact on our lives, and how we can begin to unravel the cult of productivity.

The Origins of Busyness

The glorification of busyness is deeply ingrained in our culture. From childhood, we are taught to prioritize achievement and accomplishment above all else. We are rewarded for our ability to multitask, to juggle countless responsibilities, and to always be "on the go." In the pursuit of success, we often sacrifice our well-being and neglect the importance of rest and stillness.

The Illusion of Productivity

Busyness masquerades as productivity, but in reality, it often leads to burnout, stress, and dissatisfaction. We fill our schedules to the brim, rushing from one task to the next, yet find ourselves feeling empty and depleted. The constant pressure to do more can leave us feeling trapped in a never-ending cycle of striving and never truly arriving at a sense of fulfillment.

Breaking Free from the Cult of Productivity

To break free from the cult of productivity, we must challenge the beliefs and societal norms that perpetuate it. We must recognize that our value as human beings is not determined by our level of busyness or productivity. True worth lies in our ability to cultivate meaningful connections, to nurture our well-being, and to live in alignment with our values.

Cultivating a Culture of Rest

Rest is not a sign of laziness; it is a vital component of a healthy, balanced life. By prioritizing rest and relaxation, we replenish our energy reserves, boost our creativity, and enhance our overall well-being. We must learn to embrace moments of stillness and quiet, allowing ourselves the space to recharge and rejuvenate.

Embracing the Slow Movement

The slow movement advocates for a more mindful and intentional approach to life. By slowing down and savoring each moment, we can reclaim our time and rediscover the beauty of simply being. Whether it's enjoying a leisurely meal, taking a leisurely walk in nature, or savoring a quiet moment of reflection, embracing the slow movement allows us to reconnect with what truly matters.

In unraveling the myth of busyness, we reclaim our power to live on our own terms—to prioritize presence over productivity, and to cultivate a deeper sense of fulfillment and meaning in our lives. By challenging the cult of productivity and embracing the value of stillness, we can create a world where busyness is no longer glorified, and true well-being is celebrated.

Embracing Stillness: A Journey into the Heart of Silence

In the cacophony of modern life, where noise and distractions abound, the concept of stillness can seem like a distant dream. Yet, within the depths of silence lies a profound source of peace, wisdom, and clarity. In this chapter, we embark on a journey into the heart of silence, exploring its transformative power and discovering how to embrace stillness in our everyday lives.

The Essence of Stillness

Stillness is not merely the absence of noise or movement; it is a state of being—an inner sanctuary where the mind finds peace and the soul finds solace. In stillness, we connect with the essence of our true selves, free from the incessant chatter of the mind and the distractions of the external world. It is in the quiet depths of stillness that we discover a profound sense of presence and aliveness.

The Path to Silence
Embracing stillness requires intention and practice. It is a journey inward—a pilgrimage to the sacred space within. Meditation, mindfulness, and contemplative practices serve as pathways to silence, guiding us through the labyrinth of the mind and into the serenity of the present moment. By cultivating awareness and cultivating presence, we gradually attune ourselves to the subtle whispers of silence.

The Dance of Silence and Sound
In the symphony of life, silence and sound are not opposing forces but harmonious partners. Just as a musical composition relies on the interplay of silence and sound to create beauty and depth, so too does our experience of life. By learning to embrace both the stillness within and the dynamic movement of the world around us, we discover a deeper appreciation for the richness of existence.

Finding Refuge in Silence
In moments of turmoil and upheaval, silence offers a sanctuary—a refuge from the chaos of the external world. It is within the quietude of silence that we find solace and strength, tapping into a wellspring of inner resilience and peace. Whether through meditation, prayer, or simply

sitting in quiet contemplation, we can retreat to the sanctuary of silence whenever we feel overwhelmed or lost.

The Wisdom of Silence

Silence is not empty; it is pregnant with wisdom and insight. In the stillness of silence, we encounter the depths of our own being, uncovering truths that transcend words and concepts. It is here, in the silent sanctuary of the heart, that we discover the answers to life's deepest questions and the guidance we seek on our journey of self-discovery.

Cultivating Stillness in Daily Life

While the pursuit of stillness may seem elusive amidst the hustle and bustle of everyday life, it is precisely in the midst of chaos that stillness becomes most valuable. By integrating moments of silence into our daily routines—whether through mindfulness practices, nature walks, or simply pausing to take a few deep breaths—we can cultivate a sense of inner calm and presence that permeates every aspect of our lives.

Embracing the Gift of Silence

In a world that often equates noise with productivity and busyness with success, embracing the gift of silence becomes an act of rebellion—an affirmation of our inherent worthiness beyond the need to constantly do and achieve. In the gentle embrace of silence, we reclaim our sovereignty, reconnecting with the essence of who we are and the boundless potential that resides within.

As we journey deeper into the heart of silence, we come to realize that stillness is not an escape from life but a return to its essence—a homecoming to the sacred space within. In embracing stillness, we discover a profound sense of peace, purpose, and presence that infuses every moment of our lives with beauty and meaning. May we continue to walk this path with open hearts and willing spirits, surrendering to the gentle embrace of silence and allowing its wisdom to illuminate our way.

The Power of Presence: Cultivating Mindfulness in Everyday Life

In a world that often pulls us in a hundred different directions, the practice of mindfulness offers a sanctuary—a refuge from the relentless chatter of the mind and the distractions of the external world. In this chapter, we delve into the transformative power of presence, exploring how the practice of mindfulness can enrich our lives, deepen our connections, and awaken us to the beauty of each moment.

Understanding Mindfulness
At its core, mindfulness is the practice of bringing our attention fully to the present moment, with openness, curiosity, and acceptance. It involves tuning into our

sensory experiences, thoughts, and emotions without judgment, allowing them to unfold with a sense of spacious awareness. Mindfulness is not about achieving a state of perfection or eliminating all distractions; rather, it is about cultivating a gentle presence that embraces whatever arises with kindness and compassion.

The Benefits of Mindfulness
The benefits of mindfulness extend far beyond the meditation cushion. Research has shown that regular mindfulness practice can have profound effects on our physical, mental, and emotional well-being. From reducing stress and anxiety to improving concentration and immune function, the practice of mindfulness offers a wide range of benefits that ripple out into every aspect of our lives. By cultivating a greater sense of awareness and presence, we become more attuned to our inner wisdom and the richness of our lived experience.

Bringing Mindfulness into Daily Life
Mindfulness is not limited to formal meditation practice; it can be integrated into every moment of our lives. Whether we're washing the dishes, walking in nature, or having a conversation with a friend, we can bring mindfulness to bear on our experience, anchoring ourselves in the present moment and savoring the richness of each passing moment. By cultivating a mindful attitude in our daily activities, we begin to see the world

with fresh eyes, opening ourselves to the beauty and wonder that surrounds us.

Cultivating Presence in Relationships
Mindfulness has the power to transform our relationships, deepening our connections with others and fostering a greater sense of empathy and understanding. By bringing mindful awareness to our interactions with loved ones, we can cultivate deeper intimacy, communication, and mutual respect. Mindful listening, in particular, is a powerful practice that allows us to truly hear and honor the experiences of others, fostering a sense of connection and belonging that transcends words.

Overcoming Challenges
While the practice of mindfulness offers many benefits, it is not without its challenges. The busy pace of modern life, coupled with the habitual patterns of the mind, can make it difficult to stay present and attentive. However, with patience, perseverance, and self-compassion, we can learn to navigate these challenges with grace and resilience. Each moment presents an opportunity to return to the present, to reawaken to the richness of our experience, and to cultivate a deeper sense of presence and aliveness.

Embracing Impermanence
One of the fundamental insights of mindfulness is the recognition of impermanence—the understanding that all

things are in a constant state of flux. By cultivating an attitude of acceptance and non-attachment, we learn to embrace the ever-changing nature of life with equanimity and grace. Rather than clinging to the past or fretting about the future, we can learn to meet each moment with an open heart and a sense of curiosity, knowing that each experience is a precious gift to be savored and appreciated.

The Practice of Loving-Kindness
Central to the practice of mindfulness is the cultivation of loving-kindness—a boundless sense of goodwill and compassion for ourselves and others. Through practices such as loving-kindness meditation, we can nurture feelings of warmth, acceptance, and connection, extending kindness and care to all beings, without exception. By cultivating an attitude of loving-kindness, we create an atmosphere of healing and wholeness that permeates every aspect of our lives, bringing greater joy, peace, and fulfillment to ourselves and the world around us.

As we journey deeper into the practice of mindfulness, we come to realize that presence is not a destination but a way of being—a gentle unfolding into the beauty and mystery of each moment. In cultivating mindfulness, we awaken to the richness of our lives, the interconnectedness of all beings, and the boundless potential that resides within. May we continue to walk this path with open hearts and curious minds, embracing each moment with gratitude and wonder, and allowing the transformative power of presence to illuminate our way.

Finding Peace in the Pause: Harnessing the Magic of Moments

In the fast-paced rhythm of modern life, the art of pausing is often overlooked—a casualty of our relentless pursuit of productivity and achievement. Yet, within the pause lies a profound source of peace, clarity, and renewal. In this chapter, we explore the transformative power of pausing, discovering how to embrace moments of stillness and harness their magic in our lives.

The Importance of Pausing
In a world that glorifies busyness and constant activity, the act of pausing can feel counterintuitive. However, it is in the pause that we find refuge—a sanctuary from the noise and chaos of the external world. Pausing allows us to catch our breath, recalibrate our priorities, and reconnect

with the present moment. It is in the pause that we discover the true richness and depth of life.

The Practice of Mindful Pausing
Mindful pausing involves intentionally taking breaks throughout the day to ground ourselves in the present moment and check in with our thoughts, feelings, and sensations. Whether it's a few deep breaths before a meeting, a moment of quiet reflection before starting a task, or a brief walk in nature during a busy day, mindful pausing offers an opportunity to step out of autopilot mode and reconnect with our innermost selves.

Embracing Stillness
In the pause, we encounter the magic of stillness—a timeless realm where the mind finds peace and the soul finds solace. Stillness is not the absence of noise or movement, but rather a state of being—an inner sanctuary where we can rest in the fullness of our own presence. By embracing moments of stillness throughout our day, we cultivate a sense of inner calm and spaciousness that infuses every aspect of our lives with grace and ease.

The Art of Doing Nothing
In a culture that values productivity and achievement above all else, the idea of doing nothing can feel unsettling. However, there is great wisdom in the art of doing nothing—of simply being with ourselves and the

world around us without the need for constant doing or striving. In the pause, we give ourselves permission to rest, recharge, and replenish our energy reserves, allowing us to return to our tasks with renewed clarity and vigor.

Savoring the Present Moment
Pausing allows us to savor the present moment—to fully immerse ourselves in the sights, sounds, and sensations of our experience. Whether it's enjoying a cup of tea, watching the sunset, or spending time with loved ones, pausing invites us to slow down and appreciate the beauty and wonder of life unfolding in real-time. By savoring the present moment, we cultivate a deep sense of gratitude and awe for the miracle of existence.

Cultivating Presence in Relationships
Pausing is not only essential for our own well-being but also for the health of our relationships. In the midst of busy schedules and competing priorities, taking time to pause and truly connect with others fosters deeper intimacy, empathy, and understanding. Whether it's sharing a heartfelt conversation, holding space for someone in need, or simply sitting together in companionable silence, pausing allows us to show up fully for ourselves and each other.

Overcoming Resistance
Despite the many benefits of pausing, we may encounter resistance to slowing down and taking breaks. Our inner

critic may tell us that we don't have time to pause, or that we should push through our discomfort and keep going. However, it is precisely in the moments of resistance that pausing becomes most important. By acknowledging our resistance with kindness and compassion, we can gently guide ourselves back to the present moment and rediscover the peace and clarity that await us there.

Creating Rituals of Pause
To integrate pausing into our daily lives, it can be helpful to create rituals or reminders that encourage us to slow down and take breaks. Whether it's setting aside time each day for meditation, journaling, or simply taking a few deep breaths, establishing regular pauses throughout our day helps us stay grounded and centered amidst the busyness of life. By infusing our routines with moments of stillness and reflection, we create a foundation of peace and presence that sustains us through even the busiest of days.

In the pause, we discover the beauty and magic of life unfolding in real-time. It is in these moments of stillness and reflection that we find refuge from the chaos of the world and reconnect with the essence of our true selves. May we continue to embrace the power of pausing in our lives, allowing its transformative grace to guide us on our journey of self-discovery and awakening.

Letting Go of Guilt: Liberating Yourself from the Need to Constantly Do

In a culture that values productivity and achievement above all else, the burden of guilt weighs heavily on those who dare to pause, rest, or simply be. Yet, beneath the surface of guilt lies a profound opportunity for liberation—a chance to break free from the shackles of societal expectations and reclaim our inherent worthiness. In this chapter, we explore the transformative power of letting go of guilt, discovering how to liberate ourselves from the need to constantly do and embrace the beauty of simply being.

The Guilt of Not Doing

From a young age, we are taught that our worth is tied to our productivity—that our value as human beings is determined by our ability to accomplish, achieve, and excel. As a result, the mere thought of pausing or taking a break can evoke feelings of guilt and self-doubt. We fear that if we're not constantly doing something, we're somehow falling short or failing to live up to our full potential.

The Myth of Perpetual Productivity

The myth of perpetual productivity perpetuates the belief that rest is for the weak—that those who dare to pause or take breaks are lazy, unmotivated, or lacking in ambition. Yet, the reality is that rest is not a luxury but a necessity—a fundamental human need that is essential for our physical, mental, and emotional well-being. By challenging the myth of perpetual productivity, we can begin to reclaim our right to rest and prioritize our health and happiness above the relentless pursuit of success.

Embracing the Power of Rest

Rest is not a sign of weakness but a source of strength—a wellspring of renewal and rejuvenation that replenishes our energy reserves and nourishes our souls. In the pause, we discover the power of rest—the healing balm that soothes our weary minds and bodies and restores us to wholeness. By embracing the power of rest, we honor the wisdom of our bodies and listen to the gentle whispers of our souls, allowing ourselves to be guided by the rhythm of our own inner knowing.

Letting Go of Shoulds and Shouldn'ts

Guilt often arises from the shoulds and shouldn'ts—the arbitrary rules and expectations that dictate how we should live our lives and what we should prioritize. Yet, when we let go of these shoulds and shouldn'ts and embrace the freedom to live authentically and on our own terms, we discover a profound sense of liberation and empowerment. By honoring our own needs, desires, and values, we pave the way for a life that is rich in meaning, purpose, and fulfillment.

Embracing Self-Compassion

At the heart of letting go of guilt lies the practice of self-compassion—a radical act of kindness and understanding towards ourselves, especially in moments of struggle or imperfection. Self-compassion invites us to embrace our humanity—to acknowledge our limitations, forgive our mistakes, and treat ourselves with the same warmth and tenderness that we would offer to a dear friend. By cultivating self-compassion, we create a space of refuge and acceptance within ourselves, allowing guilt to dissolve in the gentle embrace of love and forgiveness.

Rewriting the Narrative

To let go of guilt, we must rewrite the narrative that tells us we are not enough unless we are constantly doing something productive. We must challenge the cultural norms and societal expectations that perpetuate the myth of perpetual productivity and prioritize rest and self-care as essential components of a healthy, balanced life. By reclaiming our right to rest and honoring our own needs and boundaries, we pave the way for a world where guilt is replaced by compassion, and self-worth is no longer tied to productivity.

The Practice of Radical Rest

Radical rest is the practice of unapologetically prioritizing rest and self-care in a world that tells us we should always be striving and achieving. It is a revolutionary act of rebellion against the cult of busyness and the myth of perpetual productivity—a declaration of our inherent worthiness and the value of simply being. By embracing radical rest, we reclaim our power to live on our own terms and create a life that is aligned with our deepest values and desires.

Cultivating a Culture of Rest

To truly let go of guilt, we must work together to cultivate a culture of rest—one that values self-care, balance, and well-being as essential components of a thriving society.

This begins with each one of us committing to prioritize rest in our own lives and setting an example for others to follow. By modeling self-compassion and self-care, we create a ripple effect that extends far beyond ourselves, transforming our communities and the world at large.

In letting go of guilt, we reclaim our power to live authentically and on our own terms. We honor the wisdom of our bodies and the whispers of our souls, prioritizing rest and self-care as essential components of a healthy, balanced life. May we continue to challenge the myth of perpetual productivity, embrace the power of radical rest, and cultivate a culture of well-being that honors the inherent worthiness of every human being.

The Joy of Solitude: Nurturing Your Relationship with Yourself

In a world filled with constant noise and distraction, the value of solitude is often overlooked. Yet, beneath the surface of solitude lies a profound opportunity for self-discovery, reflection, and inner growth. In this chapter, we explore the transformative power of solitude, discovering how to cultivate a deep and nurturing relationship with ourselves that brings joy, fulfillment, and peace.

The Beauty of Solitude

Solitude is not loneliness—it is a choice, a sanctuary, a sacred space where we can retreat from the demands of the external world and reconnect with the essence of our true selves. In solitude, we discover the beauty of our own company, the richness of our inner landscape, and the wisdom that resides within. It is in the quiet depths of solitude that we find refuge from the noise and chaos of the world, allowing ourselves the space to rest, reflect, and recharge.

Embracing Alone Time

In a culture that often equates solitude with loneliness, embracing alone time can feel counterintuitive. However, alone time is not a punishment but a gift—a precious opportunity to cultivate self-awareness, creativity, and inner peace. Whether it's taking a solo nature walk, journaling, or simply sitting in quiet contemplation, alone time invites us to turn inward and listen to the whispers of our hearts, opening ourselves to the beauty and wonder of our own presence.

Cultivating Self-Companionship

At the heart of the joy of solitude lies the practice of self-companionship—the art of befriending ourselves and nurturing a loving and supportive relationship with the person we spend the most time with ourselves. Self-companionship involves treating ourselves with kindness, compassion, and understanding, especially in moments of struggle or pain. By cultivating a deep sense of self-compassion, we create a foundation of inner strength and resilience that sustains us through life's challenges.

Finding Freedom in Solitude

Solitude offers us the freedom to be ourselves—to explore our passions, dreams, and desires without the fear of judgment or scrutiny. In solitude, we are free to dance to the rhythm of our own hearts, to express ourselves authentically, and to follow the path that resonates most deeply with our souls. By embracing solitude as a sanctuary for self-expression and self-discovery, we liberate ourselves from the constraints of external expectations and reclaim our sovereignty as sovereign beings.

The Practice of Mindful Solitude

Mindful solitude is the practice of intentionally setting aside time for solitude and using it as an opportunity for self-reflection, introspection, and self-care. Whether it's taking a silent retreat, spending a day alone in nature, or simply carving out a few moments of quiet time each day, mindful solitude allows us to deepen our connection with ourselves and cultivate a sense of inner peace and harmony. By approaching solitude with mindfulness and intention, we transform it from a mere absence of company into a powerful catalyst for personal growth and transformation.

Rediscovering Joy in Simple Pleasures

In solitude, we rediscover the joy of simple pleasures—the beauty of a sunset, the sound of birdsong, the feel of a gentle breeze on our skin. Freed from the distractions of the external world, we are able to fully immerse ourselves in the present moment, savoring the richness of each experience with childlike wonder and delight. By cultivating a sense of curiosity and openness in solitude, we infuse our lives with a deep sense of joy and gratitude for the miracle of existence.

Embracing the Wisdom of Silence

Solitude invites us to embrace the wisdom of silence—to listen to the still, small voice within and heed its guidance. In the quiet depths of solitude, we discover that answers to life's deepest questions are not found in the noise and chaos of the world, but in the silent sanctuary of the heart. By cultivating a practice of listening in solitude, we awaken to the wisdom that resides within us, empowering us to navigate life's challenges with grace, courage, and clarity.

In embracing the joy of solitude, we reclaim our power to cultivate a deep and nurturing relationship with ourselves—one that brings joy, fulfillment, and peace. May we continue to honor the beauty of our own presence, savoring the richness of each moment in solitude and nurturing the seeds of self-love and self-discovery that reside within us. Through the practice of mindful solitude, may we cultivate a life that is rich in meaning, purpose, and joy, honoring the wisdom that lies at the heart of our own solitude.

Rest as Resistance: Challenging the Tyranny of Constant Activity

In a society that glorifies busyness and productivity, the act of resting is often seen as a luxury or even a form of laziness. However, beneath the surface of constant activity lies a deep-seated imbalance—a tyranny that robs us of our well-being and disconnects us from the rhythm of life. In this chapter, we explore the radical concept of rest as resistance, discovering how embracing rest can be an act of defiance against the relentless pressure to constantly do and achieve.

The Culture of Busyness

Our culture fetishizes busyness, equating constant activity with worthiness and success. We are conditioned to believe that our value as individuals is tied to our productivity—that our worth is measured by the number of tasks we accomplish and the hours we devote to work. As a result, we push ourselves to the brink of exhaustion, sacrificing our physical, mental, and emotional well-being in pursuit of an elusive ideal of productivity.

The Myth of Perpetual Productivity

The myth of perpetual productivity perpetuates the belief that rest is for the weak—that those who prioritize self-care and downtime are somehow less capable or deserving of success. Yet, the reality is that rest is not a luxury but a fundamental human need—a biological imperative that is essential for our health and happiness. By challenging the myth of perpetual productivity, we reclaim our right to rest and resist the cultural forces that seek to exploit our time and energy for profit.

The Power of Rest

Rest is not a passive activity but a radical act of self-care and resistance. It is in the moments of rest that we replenish our energy reserves, rejuvenate our spirits, and reconnect with the wisdom of our bodies. Rest is a form of rebellion against the tyranny of constant activity—a declaration of our inherent worthiness and the value of our well-being above all else. By embracing rest as a sacred act of resistance, we reclaim our power to live on our own terms and prioritize our health and happiness above the demands of the external world.

Restoring Balance

In a world that glorifies busyness and productivity, restoring balance requires courage and conviction. It involves challenging the cultural norms and societal expectations that perpetuate the myth of perpetual productivity and prioritizing rest as an essential component of a healthy, balanced life. By reclaiming our right to rest and resisting the pressures to constantly do and achieve, we create space for joy, fulfillment, and connection to flourish.

The Practice of Radical Rest

Radical rest is the practice of unapologetically prioritizing rest and self-care in a world that tells us we should always be striving and achieving. It is a revolutionary act of rebellion against the culture of busyness and the tyranny of constant activity—a declaration of our inherent worthiness and the value of our well-being above all else. By embracing radical rest, we reclaim our power to live authentically and on our own terms, resisting the forces that seek to exploit our time and energy for profit.

Cultivating a Culture of Rest

To truly challenge the tyranny of constant activity, we must work together to cultivate a culture of rest—one that values self-care, balance, and well-being as essential components of a thriving society. This begins with each one of us committing to prioritize rest in our own lives and setting an example for others to follow. By modeling self-compassion and self-care, we create a ripple effect that extends far beyond ourselves, transforming our communities and the world at large.

In embracing rest as resistance, we reclaim our power to live on our own terms and prioritize our health and happiness above the demands of the external world. May we continue to challenge the culture of busyness and the myth of perpetual productivity, embracing rest as a radical act of self-care and rebellion against the forces that seek to exploit our time and energy for profit. Through the practice of radical rest, may we cultivate a world that values well-being above all else, honoring the inherent worthiness of every human being to rest, recharge, and thrive.

The Art of Doing Nothing: Discovering the Beauty in Non-Doing

In a world that celebrates constant activity and productivity, the concept of doing nothing may seem counterintuitive or even taboo. However, beneath the surface of non-doing lies a profound opportunity for rest, reflection, and renewal. In this chapter, we explore the transformative power of the art of doing nothing, discovering how embracing moments of stillness and non-doing can enrich our lives and deepen our sense of well-being.

Redefining Productivity

In our culture, productivity is often equated with busyness—the more tasks we accomplish, the more productive we are deemed to be. However, true productivity is not about the quantity of tasks completed but the quality of our presence and attention in each moment. In the art of doing nothing, we redefine productivity as the ability to be fully present and engaged with whatever arises, without the need for constant action or achievement.

Embracing Stillness

The art of doing nothing begins with embracing stillness—the practice of pausing and allowing ourselves to simply be with whatever is present in the moment. Stillness is not the absence of activity but a state of inner calm and receptivity that allows us to attune ourselves to the rhythm of life. By embracing moments of stillness, we cultivate a sense of peace and presence that permeates every aspect of our lives.

Surrendering to Non-Doing

Non-doing is not laziness or apathy but a conscious choice to surrender to the flow of life and trust in the inherent wisdom of the present moment. It is a radical act of self-care and self-love that allows us to release the need for control and perfectionism and embrace the beauty of imperfection. By surrendering to non-doing, we open ourselves to the magic and mystery of life, allowing it to unfold in its own time and in its own way.

Cultivating Mindfulness

The art of doing nothing is closely linked to the practice of mindfulness—the ability to be fully present and aware of our thoughts, feelings, and sensations without judgment. By cultivating mindfulness, we learn to observe the ebb and flow of our inner experience with curiosity and compassion, allowing ourselves to be with whatever arises without the need to change or fix it. In the stillness of non-doing, we discover the beauty of the present moment and the richness of our own inner landscape.

Finding Joy in Simple Pleasures

In the art of doing nothing, we rediscover the joy of simple pleasures—the beauty of a sunrise, the laughter of a child, the warmth of a cup of tea. Freed from the pressure to constantly achieve and succeed, we are able to fully immerse ourselves in the richness of life's experiences, savoring each moment with gratitude and wonder. By embracing the art of doing nothing, we cultivate a sense of joy and contentment that transcends external circumstances and enriches our lives from within.

Letting Go of Guilt

One of the greatest challenges in embracing the art of doing nothing is letting go of the guilt and self-judgment that often accompany non-doing. In a culture that values productivity above all else, taking time for rest and relaxation can feel indulgent or selfish. However, by recognizing that rest is not a luxury but a necessity, we can release the guilt and shame that prevent us from fully embracing the art of doing nothing and prioritize our well-being with compassion and kindness.

The Gift of Presence

At its core, the art of doing nothing is about presence—being fully present and engaged with life in all its beauty and complexity. By embracing moments of stillness and non-doing, we deepen our connection to ourselves, to others, and to the world around us. We discover that true fulfillment and contentment are not found in constant striving and achievement but in the simple act of being present with whatever arises, allowing life to unfold in its own time and in its own way.

In embracing the art of doing nothing, we reclaim our power to live on our own terms and prioritize our well-being above the demands of the external world. May we continue to cultivate moments of stillness and non-doing in our lives, allowing ourselves to rest, reflect, and renew with grace and ease. Through the practice of the art of doing nothing, may we discover the beauty and richness of life that awaits us in the quiet spaces in between.

Surrendering to Silence: Deepening Your Connection to Inner Wisdom

In the midst of the noise and chaos of the world, the practice of surrendering to silence offers a sanctuary—a sacred space where we can quiet the mind, open the heart, and listen to the whispers of our inner wisdom. In this chapter, we explore the transformative power of silence, discovering how embracing moments of stillness and non-doing can deepen our connection to our innermost selves and guide us on the path of self-discovery and self-realization.

The Wisdom of Silence

Silence is not empty but pregnant with wisdom and insight. In the quiet depths of silence, we encounter the essence of our true selves—the part of us that transcends the limitations of the ego and connects us to the vastness of the universe. It is in the silence that we find refuge from the noise and distractions of the external world, allowing ourselves to attune to the subtle whispers of our inner guidance.

Surrendering to Stillness

Surrendering to silence begins with embracing stillness—the practice of pausing and allowing ourselves to be fully present in the moment. Stillness is not the absence of activity but a state of inner calm and receptivity that allows us to tune into the wisdom that resides within us. By surrendering to stillness, we create space for clarity and insight to arise, guiding us on our journey of self-discovery and self-realization.

Cultivating Inner Listening

In the silence, we cultivate the art of inner listening—the ability to tune into the quiet voice of intuition and wisdom that speaks to us from the depths of our being. Inner listening is not about seeking answers outside of ourselves but about turning inward and trusting the guidance that arises from within. By quieting the mind and opening the heart, we create space for our inner wisdom to reveal itself, illuminating the path forward with clarity and grace.

Embracing the Unknown

Surrendering to silence requires courage and trust—the willingness to let go of the need for certainty and control and embrace the mystery of the unknown. In the silence, we encounter the vastness of the void—the infinite potential that lies beyond the confines of the mind. By embracing the unknown, we open ourselves to new possibilities and experiences, allowing life to unfold in its own time and in its own way.

Nurturing Self-Compassion

In the silence, we cultivate self-compassion—the ability to treat ourselves with kindness, understanding, and acceptance, especially in moments of struggle or pain. Self-compassion is not about avoiding difficult emotions or experiences but about meeting them with love and tenderness, acknowledging our humanity and embracing ourselves with unconditional acceptance. By nurturing self-compassion in the silence, we create a space of healing and transformation that allows us to grow and evolve with grace and dignity.

Deepening Connection to Self and Others

Surrendering to silence deepens our connection to ourselves, to others, and to the world around us. In the silence, we discover the interconnectedness of all beings—the shared humanity that unites us in our joys and sorrows, our triumphs and struggles. By embracing moments of stillness and non-doing, we cultivate a sense of empathy, compassion, and understanding that transcends the limitations of the ego and opens our hearts to the beauty and wonder of life.

The Practice of Surrender

Surrendering to silence is not a passive act but an active choice—a conscious decision to let go of the need for control and allow life to unfold in its own time and in its own way. It is a practice of trust and surrender, a willingness to release the grip of the ego and surrender to the wisdom of the heart. By embracing the practice of surrender, we create space for miracles to occur, allowing the beauty and magic of life to unfold with grace and ease.

In surrendering to silence, we deepen our connection to our inner wisdom and open ourselves to the beauty and wonder of life. May we continue to embrace moments of stillness and non-doing in our lives, allowing ourselves to quiet the mind, open the heart, and listen to the whispers of our inner guidance. Through the practice of surrendering to silence, may we discover the boundless wisdom and love that reside within us, guiding us on the path of self-discovery and self-realization.

Nature's Serenade: Healing and Renewal in the Natural World

In the embrace of nature, we find solace—a sanctuary where the rhythms of the earth and the whispers of the wind soothe our souls and awaken us to the beauty and wonder of life. In this chapter, we explore the transformative power of nature's serenade, discovering how immersing ourselves in the natural world can heal, renew, and nourish our bodies, minds, and spirits.

The Healing Power of Nature

Nature has long been revered as a source of healing and renewal—a sacred space where we can reconnect with the essence of our true selves and find refuge from the stresses and strains of modern life. In the embrace of nature, we discover the healing touch of the earth—the gentle caress of the breeze, the soft kiss of the sun, the soothing murmur of the river. Nature's serenade offers us a sanctuary—a place of peace and tranquility where we can rest, recharge, and replenish our energy reserves.

Reconnecting with Our Roots

In the hustle and bustle of daily life, it's easy to lose sight of our connection to the natural world—to forget that we are part of a vast and interconnected web of life. Yet, when we immerse ourselves in nature's serenade, we rediscover our roots—we remember that we are not separate from nature but deeply intertwined with it. In the embrace of the earth, we find a sense of belonging—a feeling of kinship with all living beings that fills us with a profound sense of peace and wholeness.

Awakening the Senses

Nature's serenade awakens our senses, inviting us to fully immerse ourselves in the richness and beauty of the natural world. The scent of wildflowers, the taste of fresh spring water, the feel of the earth beneath our feet—all serve to awaken us to the present moment and fill us with a sense of wonder and awe. In the embrace of nature, we become attuned to the subtle rhythms and patterns of life, opening ourselves to the infinite beauty and magic that surrounds us.

Finding Peace in the Wilderness

In the wilderness, we find peace—a refuge from the noise and distractions of the modern world, where we can escape the pressures and demands of daily life and reconnect with our innermost selves. Whether it's a solitary hike through the forest, a leisurely stroll along the beach, or a quiet moment of reflection by a mountain stream, nature's serenade offers us a sanctuary—a sacred space where we can find solace, clarity, and renewal.

Healing the Body, Mind, and Spirit

Nature's serenade has been shown to have profound healing effects on the body, mind, and spirit. Studies have found that spending time in nature can reduce stress, anxiety, and depression, lower blood pressure and heart rate, boost immune function, and improve overall well-being. Nature's serenade offers us a natural remedy for the ailments of modern life—a tonic for the body, mind, and spirit that nourishes us from the inside out.

Cultivating Gratitude and Reverence

In the embrace of nature's serenade, we cultivate a deep sense of gratitude and reverence for the earth and all its inhabitants. We come to see ourselves not as separate from nature but as integral parts of a vast and interconnected web of life. We recognize the sacredness of the earth—the beauty, the abundance, the diversity—

and vow to protect and preserve it for future generations. In the presence of nature's serenade, we find humility and awe, opening our hearts to the wonder and mystery of the natural world.

Embracing the Call of the Wild

In the embrace of nature's serenade, we hear the call of the wild—the beckoning of the mountains, the forests, the oceans, calling us to adventure, exploration, and discovery. Nature's serenade invites us to step outside our comfort zones, to embrace the unknown, and to follow the path of our hearts. Whether it's a journey to a remote wilderness area, a camping trip under the stars, or a simple walk in the park, nature's serenade offers us endless opportunities for growth, transformation, and self-discovery.

In the embrace of nature's serenade, we find healing, renewal, and transformation. May we continue to immerse ourselves in the natural world, allowing its beauty and magic to nourish and inspire us. Through the healing touch of the earth, may we find peace, joy, and wholeness, and may we always remember to honor and protect the precious gift of nature for generations to come.

Mindful Movement: Exploring Stillness Through Yoga and Tai Chi

In the ancient practices of yoga and Tai Chi, we find pathways to stillness—a journey of mindful movement that leads us into the depths of our being and connects us to the essence of life itself. In this chapter, we explore the transformative power of mindful movement, discovering how the graceful flow of yoga and the gentle precision of Tai Chi can guide us on the path of inner peace, balance, and harmony.

The Wisdom of Yoga and Tai Chi

Yoga and Tai Chi are ancient disciplines that have been practiced for centuries, originating from different parts of the world but sharing a common goal: to cultivate a deep sense of connection to the body, mind, and spirit. Rooted in mindfulness and meditation, these practices offer us tools and techniques for navigating the complexities of modern life with grace and ease.

The Art of Yoga: Union of Body, Mind, and Spirit

Yoga is a practice of union—a sacred journey of self-discovery and self-realization that integrates body, mind, and spirit. Through a combination of physical postures (asanas), breathwork (pranayama), and meditation, yoga invites us to cultivate awareness and presence in each moment, allowing us to deepen our connection to ourselves and the world around us. In the graceful flow of yoga, we find stillness—a sanctuary where we can quiet the mind, open the heart, and awaken to the beauty and wonder of life.

The Grace of Tai Chi: Cultivating Balance and Harmony

Tai Chi is a practice of harmony—a gentle yet powerful movement meditation that cultivates balance, flexibility, and inner strength. Rooted in the principles of Taoism, Tai Chi invites us to surrender to the flow of life, embracing the natural rhythms and cycles of existence. Through slow, deliberate movements and mindful breathwork, Tai Chi harmonizes the body, mind, and spirit, allowing us to release tension and stress and find peace and serenity in the present moment.

Cultivating Mindfulness in Motion

In both yoga and Tai Chi, mindfulness is the cornerstone of practice. Mindful movement involves bringing our full attention and awareness to the present moment, allowing us to tune into the sensations of the body, the rhythm of the breath, and the quality of the mind. By cultivating mindfulness in motion, we deepen our connection to ourselves and the world around us, finding stillness amidst the movement and chaos of daily life.

Embracing the Journey Within

In the practice of yoga and Tai Chi, we embark on a journey of self-discovery—a pilgrimage into the depths of our being where we encounter the essence of who we truly are. Through the gentle guidance of these ancient practices, we learn to listen to the wisdom of the body, quiet the chatter of the mind, and connect to the source of inner peace and wisdom that resides within us. In the stillness of mindful movement, we discover the beauty and wonder of our own presence, awakening to the infinite possibilities that lie within.

The Healing Power of Presence

Yoga and Tai Chi offer us a path to healing and transformation—a sacred journey of self-healing and self-realization that nourishes the body, mind, and spirit. Through the practice of mindful movement, we release tension and stress, cultivate balance and harmony, and awaken to the beauty and wonder of life. In the healing power of presence, we find solace—a sanctuary where we can rest, rejuvenate, and reconnect to the essence of our true selves.

Integrating Mindful Movement into Daily Life

The beauty of yoga and Tai Chi lies in their accessibility—they can be practiced anytime, anywhere, by anyone. Whether it's a gentle yoga flow in the morning, a Tai Chi practice in the park at sunset, or a few moments of mindful movement during a busy day, integrating these practices into our daily lives offers us a pathway to peace, balance, and well-being. Through the practice of mindful movement, we cultivate a sense of presence and mindfulness that infuses every aspect of our lives with grace and ease.

In the graceful flow of yoga and the gentle precision of Tai Chi, we discover pathways to stillness—a journey of mindful movement that leads us into the depths of our being and connects us to the essence of life itself. May we continue to embrace these ancient practices, allowing their wisdom and grace to guide us on the path of inner peace, balance, and harmony. Through the transformative power of mindful movement, may we find stillness amidst the movement and chaos of daily life, awakening to the beauty and wonder of our own presence and the infinite possibilities that lie within.

The Beauty of Boredom: Embracing Creative Lulls and Restorative Rest

In a world that values constant stimulation and productivity, boredom is often viewed as an enemy to be avoided at all costs. However, beneath the surface of boredom lies a profound opportunity for rest, renewal, and creativity. In this chapter, we explore the transformative power of embracing boredom, discovering how moments of stillness and non-doing can spark inspiration, ignite imagination, and nourish the soul.

Rethinking Boredom

Boredom is not a lack of stimulation but a gateway to creativity—a fertile ground where new ideas can take root and flourish. In the space of boredom, we are freed from the distractions of the external world and invited to turn inward, tapping into the wellspring of creativity and imagination that resides within us. By rethinking boredom as a gift rather than a curse, we open ourselves to the infinite possibilities that lie beyond the surface of our discomfort.

Embracing Creative Lulls

Creative lulls are a natural part of the creative process—a necessary pause that allows us to recharge and replenish our creative energy. Rather than resisting these periods of stillness and non-doing, we can embrace them as opportunities for restorative rest and reflection. In the silence of creative lulls, we create space for new ideas to emerge, allowing ourselves to be guided by the rhythm of our own inner knowing.

Cultivating Presence in the Moment

Boredom invites us to cultivate presence in the moment—to fully immerse ourselves in the richness and beauty of the present experience. Whether it's staring out the window, taking a leisurely walk, or simply sitting in quiet contemplation, boredom encourages us to slow down, savor the moment, and connect to the depth and beauty of life that surrounds us. By embracing boredom as a portal to presence, we discover the richness and wonder of each moment, awakening to the beauty and magic of the world around us.

Nurturing the Imagination

In the space of boredom, the imagination is set free to wander—to explore new worlds, dream up new ideas, and envision new possibilities. Boredom ignites the fires of creativity, sparking inspiration and innovation in the most unexpected moments. By nurturing the imagination in moments of stillness and non-doing, we tap into the boundless reservoir of creativity that lies within us, allowing our dreams to take flight and our visions to become reality.

Restorative Rest and Renewal

Boredom offers us restorative rest—a chance to recharge our batteries and replenish our energy reserves. In the absence of external stimulation, we are free to rest and rejuvenate, allowing ourselves to sink into the quiet depths of relaxation and renewal. Restorative rest nourishes the body, mind, and spirit, allowing us to return to our creative endeavors with clarity, focus, and renewed vigor.

Cultivating a Culture of Rest

To truly embrace the beauty of boredom, we must work together to cultivate a culture of rest—one that values stillness, non-doing, and creativity as essential components of a balanced and fulfilling life. This begins with each one of us committing to prioritize rest and relaxation in our own lives and setting an example for others to follow. By modeling self-care and self-compassion, we create a ripple effect that extends far beyond ourselves, transforming our communities and the world at large.

In embracing the beauty of boredom, we reclaim our power to rest, recharge, and renew. May we continue to embrace moments of stillness and non-doing in our lives, allowing ourselves to sink into the quiet depths of relaxation and reflection. Through the transformative power of boredom, may we spark inspiration, ignite imagination, and nourish the soul, awakening to the beauty and wonder of life that surrounds us.

Sacred Spaces: Creating Sanctuaries for Stillness in Your Home and Mind

In the hustle and bustle of modern life, creating sacred spaces for stillness is essential for our well-being and sanity. These sanctuaries, whether physical or mental, serve as havens where we can retreat from the chaos of the world and reconnect with our inner peace and wisdom. In this chapter, we explore the transformative power of sacred spaces, discovering how intentional design and mindful practices can foster a sense of tranquility and serenity in both our homes and minds.

The Importance of Sacred Spaces

Sacred spaces are essential for our mental, emotional, and spiritual well-being. They provide us with a refuge—a sanctuary where we can rest, recharge, and replenish our energy reserves. Whether it's a cozy corner in our home, a quiet spot in nature, or a sacred ritual that we practice daily, these spaces serve as anchors in the storm of life, grounding us in the present moment and guiding us on the path of inner peace and fulfillment.

Creating Physical Sanctuaries in Your Home

Designing physical sanctuaries in our homes is a powerful way to cultivate a sense of stillness and tranquility in our lives. Whether it's a meditation corner, a reading nook, or a sacred altar, these spaces serve as reminders to slow down, breathe, and reconnect with ourselves. By intentionally curating our living spaces with elements that inspire calm and serenity, we create an environment that supports our well-being and nurtures our souls.

Designing Mental Sanctuaries in Your Mind

In addition to physical sanctuaries, we can also create mental sanctuaries in our minds—inner landscapes where we can retreat from the noise and distractions of the world and find solace in the quiet depths of our being. Through mindfulness and meditation practices, we can cultivate a sense of stillness and presence that transcends external circumstances and connects us to the essence of our true selves. By nurturing our inner sanctuaries with love and compassion, we create a foundation of inner peace and resilience that sustains us through life's challenges.

Incorporating Ritual and Ceremony

Ritual and ceremony are powerful tools for creating sacred spaces—whether physical or mental. By infusing our daily lives with intentional practices that honor the sacredness of life, we deepen our connection to ourselves, to others, and to the world around us. Whether it's lighting a candle, reciting a mantra, or simply taking a moment to breathe, these rituals serve as anchors in the present moment, guiding us back to our center and reminding us of the beauty and wonder of life.

Cultivating Presence and Stillness

Ultimately, the essence of sacred spaces lies in presence and stillness—the ability to be fully present and engaged with whatever arises, without the need for constant activity or distraction. Whether we are in our physical sanctuaries or our mental sanctuaries, the practice of presence and stillness invites us to surrender to the present moment, allowing ourselves to be held by the quiet embrace of our own being. In the stillness, we discover the beauty and wonder of life that awaits us in every moment.

In creating sacred spaces for stillness in our homes and minds, we reclaim our power to rest, recharge, and renew. May we continue to cultivate these sanctuaries with love and intention, allowing them to serve as anchors in the storm of life and guides on the path of inner peace and fulfillment. Through the transformative power of sacred spaces, may we awaken to the beauty and wonder of life that surrounds us, finding solace and sanctuary in the quiet depths of our own being.

Letting Time Unfold: Trusting the Flow of Life's Rhythms

In the relentless march of time, there is a beauty in letting go—of surrendering to the natural ebb and flow of life's rhythms. When we release the grip of control and trust in the unfolding of time, we open ourselves to a deeper sense of peace, fulfillment, and connection to the present moment. In this expansive chapter, we delve into the profound wisdom of letting time unfold, exploring how embracing the flow of life's rhythms can lead us to a more meaningful and enriching existence.

Embracing the Present Moment

At the heart of letting time unfold is the practice of embracing the present moment—the here and now where life unfolds in all its richness and complexity. By anchoring ourselves in the present moment, we release the grip of the past and future, allowing ourselves to fully experience the beauty and wonder of life as it arises. In the stillness of the present moment, we discover a profound sense of peace and presence that transcends the limitations of time and space.

Releasing the Grip of Control

Letting time unfold requires us to release the grip of control—the illusion that we can dictate the course of our lives with precision and certainty. When we relinquish the need for control and surrender to the natural flow of life, we open ourselves to new possibilities and experiences that lie beyond the confines of our expectations. In the surrender, we discover a freedom and liberation that allows us to fully embrace the journey of life with an open heart and mind.

Trusting in Divine Timing

At the heart of letting time unfold is the trust in divine timing—the belief that everything happens in its own time and in its own way, according to a higher plan that is beyond our comprehension. When we trust in divine timing, we surrender to the wisdom of the universe and allow ourselves to be guided by the flow of life's rhythms. In the trust, we find solace—a knowing that we are held and supported by a force greater than ourselves, guiding us on the path of our highest good.

Cultivating Patience and Resilience

Letting time unfold cultivates patience and resilience—the ability to wait with grace and fortitude in the face of uncertainty and adversity. When we trust in the unfolding of time, we learn to embrace the challenges and setbacks of life as opportunities for growth and transformation. In the patience, we find strength—a resilience that allows us to weather life's storms with courage and grace, knowing that every experience has its purpose and its place in the grand tapestry of existence.

Finding Beauty in the Unfolding

In the unfolding of time, we discover a beauty that transcends words—the beauty of impermanence, of change, of growth, and of transformation. When we let time unfold, we open ourselves to the infinite possibilities that lie beyond the horizon, allowing ourselves to be swept away by the currents of life's rhythms. In the beauty, we find wonder—a sense of awe and reverence for the mystery and magic of existence that surrounds us in every moment.

Living in Alignment with Life's Rhythms

Letting time unfold invites us to live in alignment with life's rhythms—to honor the cycles of birth, growth, decay, and renewal that are woven into the fabric of existence. When we attune ourselves to the natural flow of life, we find harmony—a sense of balance and wholeness that permeates every aspect of our being. In the alignment, we find fulfillment—a deep sense of purpose and meaning that arises from living in harmony with the greater unfolding of life.

In letting time unfold, we discover a profound wisdom that transcends words—a wisdom that reminds us of the beauty and wonder of existence in all its forms. May we continue to embrace the flow of life's rhythms, trusting in the unfolding of time and allowing ourselves to be swept away by the currents of existence. Through the practice of letting time unfold, may we awaken to the beauty and wonder of life that surrounds us in every moment, finding solace and sanctuary in the natural ebb and flow of existence.

The Slow Revolution: Rediscovering the Lost Art of Patience

In a world that moves at an ever-increasing pace, the art of patience is often overlooked and undervalued. Yet, beneath the surface of constant busyness lies a profound wisdom—a reminder that true fulfillment and contentment cannot be rushed or hurried, but must be savored and cherished in the present moment. In this expanded chapter, we explore the transformative power of rediscovering the lost art of patience, delving into its profound implications for our well-being, relationships, and the world at large.

The Need for a Slow Revolution
In a culture that values speed and efficiency above all else, the need for a slow revolution has never been more

pressing. The relentless pursuit of productivity and progress has left many of us feeling exhausted, disconnected, and out of balance. Yet, in our quest for more, we have overlooked the simple pleasures and joys that lie in the art of patience—the ability to slow down, savor the moment, and appreciate the beauty and wonder of life as it unfolds.

Cultivating Patience in a Fast-Paced World
Cultivating patience in a fast-paced world begins with a shift in perspective—a willingness to embrace the present moment and let go of the need for constant striving and achievement. It involves learning to trust in the natural rhythms of life, surrendering to the flow of time, and allowing ourselves to be guided by the wisdom of patience. By cultivating patience, we reclaim our power to live with intention and purpose, embracing the journey of life with an open heart and mind.

Finding Joy in the Journey
At the heart of patience lies the recognition that true fulfillment and contentment are found not in the destination, but in the journey itself. When we slow down and savor the moment, we discover a richness and depth to life that cannot be found in the pursuit of external success or validation. In the joy of the journey, we find freedom—a freedom to be fully ourselves, to follow our

hearts, and to embrace the beauty and wonder of life in all its forms.

Nurturing Connection and Presence
Patience nurtures connection and presence—the ability to be fully present and engaged with ourselves, with others, and with the world around us. When we slow down and listen deeply, we cultivate empathy, compassion, and understanding, fostering deeper connections and relationships with those around us. In the presence of patience, we discover a sense of belonging—a feeling of interconnectedness and unity that transcends the barriers of time and space.

Embracing Imperfection and Growth
In the practice of patience, we learn to embrace imperfection and growth—the understanding that progress takes time and that failure is a natural part of the journey. When we release the need for perfection and allow ourselves to make mistakes, we create space for growth and transformation to occur. In the embrace of patience, we find resilience—a resilience that allows us to bounce back from setbacks and challenges with grace and determination.

Cultivating Gratitude and Appreciation
Patience cultivates gratitude and appreciation—the ability to find beauty and wonder in the simplest of moments.

When we slow down and savor the small pleasures of life, we discover a sense of awe and reverence for the world around us. In the gratitude of patience, we find abundance—a recognition that life is filled with blessings and opportunities for joy, if only we take the time to notice and appreciate them.

The Ripple Effect of Patience
The practice of patience has a ripple effect that extends far beyond ourselves, influencing our relationships, communities, and the world at large. When we embody patience in our interactions with others, we foster deeper connections and understanding, creating a culture of empathy and compassion that transcends boundaries and differences. In the ripple effect of patience, we find hope—a hope for a world where kindness, understanding, and respect are the guiding principles of human interaction.

In rediscovering the lost art of patience, we reclaim our power to live with intention, purpose, and meaning. May we continue to embrace the slow revolution, allowing ourselves to slow down, savor the moment, and appreciate the beauty and wonder of life as it unfolds. Through the transformative power of patience, may we cultivate empathy, compassion, and understanding in our relationships and communities, fostering a world where kindness and connection are valued above all else.

The Wisdom of Non-Attachment: Releasing the Need for Constant Distraction

In a world filled with constant distractions and temptations, the wisdom of non-attachment offers us a pathway to inner peace, freedom, and fulfillment. Non-attachment is not about detachment or indifference, but rather a deep understanding that our sense of self-worth and happiness does not depend on external circumstances or possessions. In this expanded chapter, we explore the transformative power of releasing the need for constant distraction, discovering how cultivating non-attachment can lead us to a deeper sense of contentment and joy.

Understanding Non-Attachment

Non-attachment is rooted in the recognition that true happiness and fulfillment cannot be found in the pursuit of external possessions or achievements. It is a state of being in which we release the grip of attachment to outcomes, expectations, and desires, and instead embrace the present moment with acceptance, openness, and equanimity. Non-attachment does not mean denying our desires or ambitions, but rather cultivating a sense of inner peace and contentment that is not dependent on external conditions.

Letting Go of the Ego

At the heart of non-attachment lies the willingness to let go of the ego—the part of us that seeks validation, approval, and control from the outside world. When we release the need to constantly feed the ego with external praise or possessions, we free ourselves from the cycle of craving and aversion that leads to suffering. In the surrender of the ego, we discover a sense of freedom—a freedom to be fully ourselves, without the need for validation or approval from others.

Embracing Impermanence

Non-attachment invites us to embrace impermanence—the understanding that all things in life are transient and ever-changing. When we let go of the need to cling to things or experiences, we open ourselves to the flow of life's rhythms, allowing ourselves to be fully present and engaged with whatever arises. In the embrace of impermanence, we find liberation—a liberation that allows us to fully experience the beauty and wonder of life in all its forms.

Cultivating Contentment and Joy

Non-attachment cultivates contentment and joy—the ability to find happiness and fulfillment in the present moment, regardless of external circumstances. When we release the need for constant distraction and busyness, we create space for gratitude, appreciation, and wonder to arise. In the contentment of non-attachment, we find peace—a peace that transcends the ups and downs of life and allows us to rest in the quiet depths of our own being.

Letting Silence Speak

In the practice of non-attachment, we learn to let silence speak—to quiet the chatter of the mind and listen deeply to the wisdom that arises from within. When we release the need for constant stimulation and distraction, we create space for insight, clarity, and intuition to emerge. In the silence of non-attachment, we find wisdom—a wisdom that guides us on the path of self-discovery and self-realization.

Cultivating Non-Attachment in Daily Life

Cultivating non-attachment in daily life involves mindfulness and awareness of our thoughts, feelings, and actions. It requires us to notice when attachment arises and to gently let it go, returning to the present moment with acceptance and equanimity. Through practices such as meditation, mindfulness, and self-inquiry, we can cultivate a sense of non-attachment that permeates every aspect of our lives, leading us to greater peace, freedom, and fulfillment.

The Liberation of Non-Attachment

In the liberation of non-attachment, we find freedom—a freedom to be fully ourselves, without the need for external validation or approval. We release the grip of attachment to outcomes, expectations, and desires, and instead embrace the present moment with acceptance, openness, and equanimity. In the wisdom of non-attachment, we discover a profound sense of peace, joy, and contentment that arises from within, independent of external circumstances or possessions.

In embracing the wisdom of non-attachment, we reclaim our power to live with freedom, peace, and fulfillment. May we continue to cultivate non-attachment in our lives, releasing the need for constant distraction and finding joy and contentment in the present moment. Through the transformative power of non-attachment, may we discover a deeper sense of peace, freedom, and fulfillment that arises from within, guiding us on the path of self-discovery and self-realization.

The Practice of Deep Listening: Cultivating Presence in Relationships

In our fast-paced and often distracted world, the practice of deep listening offers a profound antidote to the disconnection and misunderstandings that can arise in our relationships. Deep listening is more than just hearing words—it is about being fully present and attentive to the speaker, with an open heart and mind. In this expanded chapter, we explore the transformative power of cultivating deep listening in our relationships, discovering how it fosters understanding, empathy, and connection.

The Essence of Deep Listening

Deep listening is rooted in the intention to fully understand and connect with the speaker, without judgment or interruption. It involves giving our undivided attention to the speaker, tuning in to both their words and their underlying emotions and intentions. Deep listening requires us to set aside our own agenda and opinions, and to be fully present with the speaker in the moment.

Creating Space for Presence

In the practice of deep listening, we create a sacred space for presence—a space where the speaker feels heard, understood, and valued. This requires us to set aside distractions and preconceived notions, and to be fully present with the speaker in body, mind, and spirit. By creating a space of presence, we invite the speaker to open up and share their thoughts, feelings, and experiences more freely.

Cultivating Empathy and Understanding

Deep listening fosters empathy and understanding—the ability to see the world from the speaker's perspective and to validate their experiences and emotions. By listening deeply, we cultivate empathy and compassion for the speaker, acknowledging their feelings and struggles without judgment or criticism. In the practice of deep listening, we bridge the gap between ourselves and others, fostering deeper connections and relationships.

Honoring the Power of Silence

In deep listening, we honor the power of silence—the space between words where deeper truths and insights can emerge. Silence allows us to process and reflect on what has been said, and to respond with thoughtfulness and care. By embracing silence, we create space for deeper connection and understanding to unfold, allowing the speaker to feel truly heard and valued.

Overcoming Barriers to Listening

In our relationships, we may encounter barriers to deep listening, such as distractions, defensiveness, or the desire to fix or solve the speaker's problems. By recognizing and overcoming these barriers, we can create a more supportive and nurturing environment for communication and connection. This involves practicing mindfulness and self-awareness, and cultivating a willingness to let go of our own agenda and ego in service of the speaker's needs.

Deepening Connection and Trust

Deep listening deepens connection and trust—the foundation of healthy and fulfilling relationships. When we listen deeply to others, we create a bond of trust and understanding that strengthens our relationships and allows them to thrive. By practicing deep listening, we create a space for vulnerability and authenticity, where we can truly be ourselves and feel accepted and valued by others.

Integrating Deep Listening into Daily Life

Deep listening is a practice that can be integrated into every aspect of our lives, from our interactions with loved ones to our conversations with colleagues and acquaintances. By bringing mindfulness and presence to our communication, we can foster deeper connections and understanding in all of our relationships. Through the practice of deep listening, we cultivate a more compassionate and empathetic way of being in the world, enriching our lives and the lives of those around us.

In embracing the practice of deep listening, we unlock the transformative power of presence, empathy, and understanding in our relationships. May we continue to cultivate deep listening in our lives, creating spaces of presence and connection where all voices are heard and valued. Through the practice of deep listening, may we foster deeper connections and understanding in our relationships, and create a more compassionate and empathetic world for ourselves and future generations.

Restoring Balance: Nourishing Your Body, Mind, and Spirit

In the hustle and bustle of modern life, it's easy to lose sight of the importance of balance—of nurturing not just our physical health, but also our mental and spiritual well-being. Yet, true vitality and fulfillment come from aligning all aspects of ourselves—body, mind, and spirit—in harmony with each other. In this expanded chapter, we delve into the profound wisdom of restoring balance, exploring how we can nourish ourselves holistically to cultivate a life of vitality, joy, and purpose.

The Importance of Balance

Balance is the cornerstone of health and well-being—it is the foundation upon which vitality and fulfillment are built. When our lives are out of balance, we may experience physical ailments, mental stress, and spiritual disconnection. Yet, by nurturing all aspects of ourselves—body, mind, and spirit—we can restore harmony and wholeness to our lives, allowing us to thrive in every aspect of our being.

Nourishing Your Body

Nourishing your body is essential for vitality and well-being. This involves eating a balanced diet rich in whole, nutrient-dense foods, staying hydrated, and getting regular exercise. It also means listening to your body's needs and honoring its signals for rest, relaxation, and rejuvenation. By nourishing your body with love and care, you cultivate a strong foundation for health and vitality that supports you in living your fullest life.

Cultivating Mental Well-being

Cultivating mental well-being is vital for a balanced and fulfilling life. This involves practicing mindfulness and self-awareness, managing stress, and fostering positive thoughts and emotions. It also means engaging in activities that bring you joy and fulfillment, such as creative expression, hobbies, and spending time in nature. By nurturing your mental well-being, you cultivate a sense of peace and clarity that allows you to navigate life's challenges with resilience and grace.

Connecting with Your Spirit

Connecting with your spirit is essential for a sense of purpose and fulfillment. This involves exploring your beliefs and values, engaging in spiritual practices that resonate with you, and fostering a sense of connection to something greater than yourself. It also means listening to your intuition and following your heart's guidance in all areas of your life. By nurturing your spirit, you cultivate a deep sense of purpose and meaning that infuses every aspect of your being with joy and vitality.

Integrating Holistic Practices

Integrating holistic practices into your daily life is key to restoring balance and harmony. This may include yoga, meditation, breathwork, energy healing, or any other practices that nourish your body, mind, and spirit. By incorporating these practices into your routine, you create space for self-care and self-discovery, allowing yourself to connect more deeply with your inner wisdom and intuition.

Embracing Self-Care

Embracing self-care is essential for restoring balance and vitality. This involves prioritizing your own needs and making time for activities that nourish and rejuvenate you. Whether it's taking a relaxing bath, spending time in nature, or simply curling up with a good book, self-care allows you to replenish your energy reserves and cultivate a sense of well-being that radiates from the inside out.

Cultivating Gratitude and Joy

Cultivating gratitude and joy is a powerful way to restore balance and harmony in your life. This involves taking time each day to acknowledge and appreciate the blessings in your life, no matter how small. It also means finding joy in the present moment, whether it's through laughter, connection with loved ones, or simply being fully present with whatever arises. By cultivating gratitude and joy, you shift your focus from what's lacking to what's abundant, allowing yourself to experience greater fulfillment and contentment in every aspect of your life.

In restoring balance and harmony to your life, you reclaim your power to live with vitality, joy, and purpose. May you continue to nourish your body, mind, and spirit with love and care, cultivating a life of holistic well-being and fulfillment. Through the practice of restoring balance, may you awaken to the beauty and wonder of life that surrounds you, embracing each moment with gratitude, joy, and reverence.

The Art of Savoring: Finding Delight in Life's Simple Pleasures

In a world filled with constant busyness and distractions, the art of savoring offers us a powerful reminder to slow down, be present, and appreciate the richness and beauty of life that surrounds us. Savoring is more than just enjoying the moment—it's about fully immersing ourselves in the experience, engaging all our senses, and allowing ourselves to be swept away by the wonder and delight of the present moment. In this expanded chapter, we delve into the transformative power of savoring, exploring how it can bring greater joy, fulfillment, and meaning into our lives.

The Essence of Savoring

Savoring is the practice of fully engaging with and enjoying the present moment, with all its sights, sounds, smells, tastes, and textures. It involves slowing down and being fully present, allowing ourselves to soak up the richness and beauty of life that surrounds us. Savoring is about cultivating a sense of wonder and appreciation for the simple pleasures that bring us joy and fulfillment.

Engaging the Senses

At the heart of savoring lies the engagement of the senses—the ability to fully experience the richness and depth of each moment. Whether it's savoring a delicious meal, listening to music, watching a sunset, or feeling the warmth of the sun on your skin, savoring invites us to engage all our senses and immerse ourselves fully in the experience. By engaging the senses, we awaken to the beauty and wonder of life that surrounds us in every moment.

Cultivating Mindfulness

Savoring is a form of mindfulness—a practice of paying attention to the present moment with openness, curiosity, and acceptance. It involves letting go of distractions and preoccupations, and instead focusing our attention on the here and now. By cultivating mindfulness through the practice of savoring, we deepen our connection to ourselves, to others, and to the world around us, fostering greater presence, awareness, and appreciation in our lives.

Finding Joy in the Ordinary

Savoring invites us to find joy in the ordinary moments of life—to appreciate the simple pleasures that bring us happiness and fulfillment. Whether it's a cup of coffee in the morning, a walk in nature, or a quiet moment spent with loved ones, savoring reminds us that joy can be found in the smallest of moments. By savoring the ordinary, we cultivate a sense of gratitude and contentment that permeates every aspect of our lives.

Creating Rituals of Savoring

Rituals of savoring are powerful tools for cultivating presence and appreciation in our lives. Whether it's setting aside time each day to savor a cup of tea, taking a leisurely walk in nature, or practicing gratitude before bedtime, rituals of savoring help us to slow down, be present, and connect more deeply with the richness and beauty of life that surrounds us. By creating rituals of savoring, we infuse our daily lives with moments of joy, wonder, and delight.

Embracing the Transience of Experience

Savoring also involves embracing the transience of experience—the understanding that all moments, no matter how beautiful or joyful, are fleeting and impermanent. By savoring the present moment, we acknowledge the preciousness of life and the need to fully appreciate and cherish each moment as it arises. In embracing the transience of experience, we deepen our appreciation for the beauty and wonder of life, and cultivate a sense of gratitude for the gift of each moment.

Sharing Savoring with Others

Savoring is not meant to be a solitary practice—it's something that can be shared and enjoyed with others. Whether it's savoring a meal together, watching a sunset with a loved one, or simply sharing stories and laughter, savoring deepens our connections and relationships with others, fostering greater intimacy, empathy, and understanding. By sharing savoring with others, we create moments of joy and connection that enrich our lives and nourish our souls.

In embracing the art of savoring, we reclaim our power to fully engage with and appreciate the richness and beauty of life that surrounds us. May we continue to cultivate moments of savoring in our daily lives, allowing ourselves to slow down, be present, and immerse ourselves fully in the wonder and delight of each moment. Through the practice of savoring, may we awaken to the beauty and wonder of life that surrounds us in every moment, finding joy, fulfillment, and meaning in the simple pleasures that bring us happiness and contentment.

Embracing the Journey: Sustaining Your Practice of Being Still

In the midst of life's chaos and demands, the practice of being still offers us a sanctuary—a place of refuge where we can find peace, clarity, and renewal. Yet, sustaining this practice requires commitment, perseverance, and a willingness to embrace the journey of self-discovery and growth. In this expanded chapter, we delve into the transformative power of embracing the journey of being still, exploring how it can deepen our connection to ourselves, others, and the world around us.

The Continual Journey of Being Still

Being still is not a destination, but rather a journey—a continual unfolding of self-discovery, growth, and transformation. It's about learning to quiet the mind, open the heart, and connect more deeply with the essence of who we are. The journey of being still is an ongoing process of exploration and discovery, where each moment offers new insights, revelations, and opportunities for growth.

Cultivating Patience and Perseverance

Sustaining your practice of being still requires patience and perseverance. It's about showing up for yourself each day, even when the mind is restless or the demands of life feel overwhelming. By cultivating patience and perseverance, you create a strong foundation for your practice, allowing yourself to weather the challenges and setbacks that inevitably arise along the way.

Nurturing Self-Compassion and Kindness

Being still is not about perfection—it's about showing up with kindness and compassion for yourself, exactly as you are. It's about embracing your humanness and allowing yourself to be gentle with yourself when you stumble or fall. By nurturing self-compassion and kindness, you create a supportive and nurturing environment for your practice to flourish, allowing yourself to learn and grow from every experience.

Embracing Resistance and Discomfort

Resistance and discomfort are natural parts of the journey of being still—they are the signposts that indicate where growth and transformation are occurring. Instead of avoiding or pushing away these feelings, embrace them with curiosity and openness. Explore what lies beneath the surface, and allow yourself to sit with whatever arises with courage and grace.

Cultivating Presence in Everyday Life

Being still is not confined to your meditation cushion—it's about cultivating presence in every moment of your life. It's about bringing mindfulness and awareness to your daily activities, interactions, and experiences. By integrating the principles of being still into your everyday life, you create opportunities for deeper connection, meaning, and fulfillment in everything you do.

Finding Support and Community

Sustaining your practice of being still is easier when you have support and community to lean on. Seek out like-minded individuals who share your commitment to self-discovery and growth. Whether it's joining a meditation group, attending a retreat, or simply connecting with others online, find ways to nurture a sense of belonging and support in your journey.

Celebrating Your Progress and Growth

As you journey deeper into the practice of being still, take time to celebrate your progress and growth. Acknowledge the milestones you've reached, no matter how small, and honor the courage and commitment it takes to show up for yourself each day. By celebrating your progress and growth, you cultivate a sense of joy, gratitude, and appreciation for the journey you're on.

In embracing the journey of being still, you open yourself to a world of possibility and potential. May you continue to nurture your practice with patience, perseverance, and kindness, allowing yourself to embrace the journey of self-discovery and growth with an open heart and mind. Through the transformative power of being still, may you awaken to the beauty and wonder of life that surrounds you in every moment, finding peace, clarity, and renewal on the path of self-discovery and growth.

The best quotes about doing nothing

A life spent making mistakes is not only more honorable, but more useful than a life spent doing nothing.

George Bernard Shaw

The only thing necessary for the triumph of evil is for good men to do nothing.
Edmund Burke

Doing something costs something. Doing nothing costs something. And, quite often, doing nothing costs a lot more!
Ben Feldman

Life is inherently risky. There is only one big risk you should avoid at all costs, and that is the risk of doing nothing.
Denis Waitley

Nothing comes from doing nothing.

William Shakespeare

The problem with doing nothing is not knowing when you're finished.

Benjamin Franklin

Don't underestimate the value of Doing Nothing, of just going along, listening to all the things you can't hear, and not bothering.

A. A. Milne

The master accomplishes more and more by doing less and less until finally he accomplishes everything by doing nothing.

Laozi

It is awfully hard work doing nothing.

Oscar Wilde

To do anything, it is first necessary to be doing nothing.

Nancy Hale

Doing nothing is better than being busy doing nothing.

Laozi

Making no decision is a decision. To do nothing. And doing nothing always brings you to nowhere.

Robin Sharma

Criticism is something we can avoid easily by saying nothing, doing nothing, and being nothing.

Aristotle

Don't underestimate the value of Doing Nothing

A. A. Milne

All that needs to be done for evil to prevail is good men doing nothing.

Edmund Burke

Some of us need to discover that we will not begin to live more fully until we have the courage to do and see and taste and experience much less than usual... And for a man who has let himself be drawn completely out of himself by his activity, nothing is more difficult than to sit still and rest, doing nothing at all. The very act of resting is the hardest and most courageous act he can perform.

Thomas Merton

Doing nothing is very hard to do... you never know when you're finished.

Leslie Nielsen

The ultimate happiness is doing nothing.

Zhuangzi

Failure should be our teacher, not our undertaker. Failure is delay, not defeat. It is a temporary detour, not a dead end. Failure is something we can avoid only by saying nothing, doing nothing, and being nothing.

Denis Waitley

Maturity is the ability to make a decision and stand by it. Immature people spend their lives exploring endless possibilities and then doing nothing. Action requires courage. Without courage, little is accomplished.

Ann Landers

This is very important -- to take leisure time. Pace is the essence. Without stopping entirely and doing nothing at all for great periods, you're gonna lose everything...just to do nothing at all, very, very important. And how many people do this in modern society? Very few. That's why they're all totally mad, frustrated, angry and hateful.

Charles Bukowski

I knew exactly what I was doing: I was doing nothing. because I knew there was nothing to do.

Charles Bukowski

I need so much time for doing nothing that I have no time for work.

Pierre Reverdy

Your reality is yours. Stop wasting time looking at someone else's reality while doing nothing about yours.

Steve Harvey

When was the last time you spent a quiet moment just doing nothing - just sitting and looking at the sea, or watching the wind blowing the tree limbs, or waves rippling on a pond, a flickering candle or children playing in the park?

Ralph Marston

www.ingramcontent.com/pod-product-compliance
Lightning Source LLC
Chambersburg PA
CBHW071214240526
45470CB00018B/1861